THE HIGH VALUE MAN:
Principles of Positive Masculinity

by Min Liu

http://www.artofverbalwar.com

© 2016 Art of Verbal War. All Rights Reserved.

May this book enable you to fully

realize all the value in you that you

never knew you had.

TABLE OF CONTENTS

Chapter 1: Introduction

The Typical Alpha Male
The Typical Beta "Male"
Enter The "High Value Man"
High Value Men Focus On The Internal, Not The External
The Benefits of Becoming a High Value Man

Chapter 2: The Six-Part Plan For Becoming A High Value Man

Part 1 – Eliminating Negative Behaviors
Part 2 – Guiding Principles
Part 3 – Value Enhancers
Part 4 – Aggressiveness vs. Assertiveness
Part 5 – The High Value Man In Practice
Part 6 – Exercises

Chapter 3: Conclusion

Conclusion
About The Author
Also by Min Liu

Want to continue your journey towards becoming a high value man?

For the truly committed only...

DOWNLOAD YOUR FREE COPY OF THE HIGH VALUE MAN CHECKLIST!

www.artofverbalwar.com/thvmbonus

CHAPTER ONE
Introduction

Introduction

We are at an inflection point for manhood.

In fact, some say we are well beyond an inflection point, and that we have already reached a point of no return.

I don't know if that's true, but there is certainly a crisis brewing. Ultimately, it doesn't matter whether we have reached a point of no return. Each one of us individually can still choose the path that is for us.

For those of you who are not familiar with the term "inflection point", it is the point of a curve at which a change in the direction of curvature occurs.

So, what is this inflection point in respect of manhood?

Simply put, at the time you are now reading this book, men have lost their way.

When I say "lost their way", I mean that most men have veered off onto two different paths in terms of their manhood and masculinity.

Everywhere you look, you will find men on one of these paths or the other of these paths, neither of which are ideal.

There is a divergence in what people believe manhood is supposed to look like.

And, not only is there a divergence in our not-so-collective image of manhood, but the two different paths that are available for men to traverse are both sub-optimal.

Many forces and environmental influences have shaped this divergent reality as it exists today. This book does not focus on what these forces are, but if you are motivated to find out, you will find many authors who write extensively about this aspect and are more knowledgeable than I am.

Instead, this book merely seeks to deal with reality as it exists today.

And, this reality is that there are currently two paths that most men follow.

One path is the path of the so-called "alpha male". This is a path that has been luring more and more men over the years ever since the concept of being an "alpha male" has come into the mainstream as a "desirable" thing for men to aspire to. You will also find people using the word "bad boy" as a proxy for alpha males, i.e. being a "bad boy" is a "good" thing.

The group of "alpha males" consists of a small group of natural alpha males, which most men are not, and a much larger group of men who are striving to become "alpha males" or "bad boys".

When I use the term "alpha male" in this book, I'm mostly referring to this latter group of men who are not natural alpha men, but men trying to become "alpha".

It is said that "alpha males" are more successful in life, whether in their careers or with women, and whenever these two rewards are promised, people will flock to them like shoppers from the dregs of society to the local Walmart's front doors on a Black Friday clamoring to get the latest electronics at an extreme bargain.

We will get into this concept further in a little bit, but these "alpha males", "bad boys", and the people who espouse the theory of "alpha males" are too caught up in the nets of social hierarchy, social dominance, and rankism ("rankism" meaning abusive, discriminatory, or exploitative behavior towards people because of their so-called rank or perceived rank in a particular hierarchy).

The problem with most of these unnatural "alpha males" is that in trying to become more "masculine", they end up reflecting many aspects of "masculinity" that are problematic. Those who espouse the concept of "becoming alpha males", while mostly well-meaning, have promoted a type of masculinity that is a "negative masculinity".

I must emphasize that many of these people are well-meaning, but either their advice is misguided or their advice is misinterpreted. In either case, the outcome is negative.

Now let's talk about the other path I referred to earlier. This other path is the path of the "beta male". This other path is the path of the "beta male" or "low value male", two terms I will use interchangeably in this book.

To be clear, the "beta male" path isn't a path that is necessarily chosen by men who are on this path. These men usually end up on this path by default, by societal programming, by virtue how they were raised, or by the different environments they have existed in, or circumstances they have experienced in life.

Or, it could also be a path explicitly chosen by these men, in order to serve as a counterpoint to alpha men and bad boys, men they have been led to believe represent all the negatives about masculinity.

"Beta males", while they do not necessarily exhibit the type of "negative" masculinity that alpha males and bad boys do, are lacking in the the "masculinity" aspect of the equation.

And, in many ways, the behaviors of "beta males" are also negative.

Beta males often shy from the aspects of masculinity that are positive, such as wielding their power for good purposes and demonstrating their integrity consistently. Often, they hide their true desires and end up later expressing them in passive-aggressive ways. And, in many ways, their behavior is an "act" just like an "alpha male's" unnatural behavior is ultimately an act.

A classic beta male pose, i.e. "The Hovering Hand"

"Power-over is all about dominating or controlling another or others, and is a key operational preference of unhealthy manhood, along with power-under (meaning find a certain status or safety in submissively aligning with more dominant or privileged males). Power-over and power-under are foundational partners in many relationships,

often constituting a cult of two - until the downtrodden one takes a real stand against the dominating one."

-Robert Augustus Masters *("To Be A Man")*

At the end of the day, neither alpha males nor beta males stand for what I call "positive masculinity".

Because of these two divergent paths, what we are now seeing is a CHASM amongst men.

We have on one side, the alpha males and on the other side, the beta males.

There are virtually no men who stand within the chasm between these two types of men, men who in this book I will refer to later on in this book as "high value men", men who I believe reflect masculinity in the most positive way possible.

These "high value men" are neither alpha males nor beta males. To the extent you can even find one of these "high value men", they simply exist in their own category of men and transcend the alpha/beta dichotomy.

As I alluded to earlier, some think we have reached a point of no return.

I don't believe this is true.

This book, **The High Value Man**, is about showing men a different path for their masculinity, and the principles to either regain their manhood or hone and transform their manhood into something completely different, and positive in every way.

The goal is to assist men in transcending the alpha/beta dichotomy, a paradigm of masculinity that was well-meaning, but shallow, underdeveloped, and misguided at its core.

Men do not need to be "alpha" or a "bad boy".

They certainly do not want to be "beta".

What you should instead aim to become is "high value".

I want to be clear. I'm definitely not the first one to come up with this general concept of a high value man. Some others have theorized about this special breed of man, perhaps using different terminology.

Instead, my goal for this book is to bring more awareness to the idea of a "high value man" and to flesh it out further so that it becomes useful and practical instead of it forever remaining a theoretical construct.

So, if you want to reclaim and redevelop masculinity from these conventional, yet false paths, especially your own masculinity, then allow me to blaze a new path for you.

To winning,
Min Liu

NEXT STEPS:

DOWNLOAD your **SPECIAL BONUS** *"The High Value Man Checklist"* **($99 price, infinitely valuable)**, containing a summary of ACTION STEPS to continue your transformation into a high value man at www.artofverbalwar.com/thvmbonus

SUBSCRIBE to my YouTube channel, The Art of Verbal War, where people learn to EXCEL at verbal skills at www.youtube.com/artofverbalwar

READ MORE about verbal skills, power, persuasion, and influence at my blog at www.artofverbalwar.com/blog

SEND ME A MESSAGE at info@artofverbalwar.com

Want to continue your journey towards becoming a high value man?

For the truly committed only...

DOWNLOAD YOUR FREE COPY OF THE HIGH VALUE MAN CHECKLIST!

www.artofverbalwar.com/thvmbonus

The Typical Alpha Male

Let's start with the "alpha male".

What does the typical "alpha male" look like?

On a recent trip to Los Angeles, these men were swarming everywhere, like vultures to a dead piece of meat.

Hyper-aggressive, out of control, douche-bagging, chest-thumping sociopathic men, or shall I say bullies?

I'm hardly one who likes to stereotype people, but a lot of these "alpha men" do nothing to dispel the notion that these many of these men are bullies, assholes, and douchebags.

Here's a quote from Robert Augustus Master's "To Be A Man" which describes these unnatural alpha males in a way I wholeheartedly agree with:

"Being a man has very little do with <u>trying</u> to be a man, and a lot to do with being present and trustworthy, grounded and transparent, and showing up as a warrior of integrity and intimacy, compassionately citing through the roots of whatever is obstructing one's well-being."

I am referring mostly not to those who are natural alpha males (who are few and far between), but those who are "acting", or shall I say trying too hard, to be more "alpha".

Here are some common traits of these so-called "alpha males": prideful, petty, unreflective, rigid, overconfident, and hyper-aggressive, i.e. so called "dominance".

And even worse still, many of these men are also manipulative and amoral. It is one thing to have an overinflated view of yourself, but another thing altogether to have that play out onto other people.

While it is most certainly a desirable thing to not be overly constrained by the opinions of others and values of society, as we will talk about later, the anti-social aspects of some alpha males are not desirable traits.

"In short, a simple dominant-nondominant dimension may be of limited value when predicting mate preferences for women."

-Jerry Burger & Mica Crosby

We will talk about balancing anti-social and pro-social traits later.

Trust me, I've been on this path before myself. Instead of being a solid, grounded man, I've been through a phase of my life where in seeking to become more "masculine", I ended up demonstrating a lot of the negative traits I write about now.

Compensating for my own insecurities and being rather uncalibrated, while also deluding myself about the so-called "confidence" I had, I became a version of myself that was ridiculous.

These "alpha males" may think very highly of themselves, but other than themselves, nobody else thinks very highly of them. To the outside world, these "alpha males" are indeed ridiculous.

Now, as with many things in life, it is not necessarily the idea, but the execution that is problematic.

Alpha and beta are social hierarchy and social dominance concepts that we as humans borrowed from the animal kingdom.

In the animal kingdom, the alpha is the highest ranked individual within an animal group. He eats and mates first. All others defer to the alpha, who is generally the strongest, fittest and most powerful male in the group.

In trying to explain human behavior (and trying to provide a useful model of masculinity for men to follow), various people have co-opted this animal kingdom concept and created a psuedo-scientific construct in an attempt to explain what male "success" looks like and/or how to obtain such "success".

Ironically, there is even evidence out there that the social structure of wolf packs, for example, do not exactly work the way we think they do.

In our minds, the human form of the alpha is often a "man" who earns a big paycheck, drives fast cars, gets to have sex with beautiful women, and commands beta men around.

At nightclubs, he parts the crowd and nabs the doting attention of the bartender with a nod. He picks up beautiful women with ease and leaves the club with one on each arm.

In business, he takes command in every meeting and dominates all others into submitting to his brilliant, incontrovertible ideas and orders.

And at social functions, he is the life of the party.

That is a seductive picture, I know, but reality doesn't always reflect our wild imaginations.

While some "alpha males" may live their lives like this, there are many aspects of these alpha males that are anti-social, counterproductive, and problematic.

Let me give you an example of an "alpha male":

My friend used to a have a boss who was the top-ranking lawyer at his company. He clearly saw himself an "alpha male".

He was a knowledgeable, experienced, and well trained lawyer. He won many legal battles through the sheer force of his personality, especially the "bullying" side of him, which in an ironic way (as you will see), served him well in his job.

But this story is not about his legal prowess.

Even though he was a full-time lawyer, he also served in the Army reserve, and held a black belt in karate.

For the purposes of telling this story, let's call him the "Black Belt Bully". He had a black belt in karate, and when it came to bullying other people, he was a full on Sensei Kreese from the move "The Karate Kid" (the original one, not the Jaden Smith one!).

I totally dated myself with that reference, but I digress…

The Black Belt Bully's office was adorned with pictures of himself in military fatigues and martial arts gear, and also certificates, awards, and trophies of his accomplishments in these areas.

The Black Belt Bully loved to talk about himself, his accomplishments, and made sure everyone knew about them. In doing so, he would frequently talk about how other people were "pussies" and not as "brave" as him. He was full of putdowns for other people, and of course, he believed he was always different and better than other people.

Once a month, he would come into the office wearing his military fatigues on days he had to report for reserve duty, even though he clearly had time to change his outfit before having to leave for duty.

Frequently, he would dress people down when they didn't meet his generally unreasonable expectations. Luckily, my friend was never on the receiving end of one of these dressdowns.

I'm certain The Black Belt Bully felt more "manly" when he did one of these dressdowns.

One day he dressed down another person on his team, who ended up in tears. After she stopped crying, she promptly reported this incident to HR.

The company CEO and HR had a talk with him to ask him to tone down his behavior and be more constructive in providing feedback to his employees. The CEO did not dress him down in this meeting, but merely requested a modification in the way he dealt with employees.

Instead of taking this discussion and the CEO's request to heart as a professional and mature man would do, the Black Belt Bully dug in and refused to admit fault. He never entertained the mere possibility that he might have been wrong in non-constructively dealing with his employee.

Instead, he kept up his offending behavior, and kept pressing the employee who reported the incident, and he started avoiding the CEO.

Within six months of that incident, the Black Belt Bully was pushed out and let go.

Last time I heard, he wasn't able to find another job, and is now a solo lawyer working out of his house.

While my friend's former boss was "dominant" as alpha males are expected to be, the Black Belt Bully was not well respected by others, but merely feared or barely tolerated.

While he was successful in some respects in his career, his "alpha male" behaviors ended up doing him in and prevents him from being employable to this day. From what I hear, his personal life is no better.

I'm no Carl Jung, but his outward "toughness" actually ended up revealing his deep insecurities. Not secure in himself enough to admit fault when he was incorrect, he ended up digging his own grave at work because he refused to take steps to correct behavior that needed to be modified.

Not being able to accept and modify one's own shortcomings is actually a sign of insecurity and weakness. It is not "beta" or a sign of weakness to understand one's own weaknesses and work on them. If you believe this, then you should know that this viewpoint is merely your own interpretation of what a "masculine" man should be.

In the animal kingdom, life is most likely to be short. Dominance plays an important role in an animal's survival, and that of its pack.

However, in the human kingdom, life is generally not short. Acting like an "alpha male" as if you are an animal in the wild is unsustainable.

There is another way for masculinity to be expressed POSITIVELY, as you will see later.

"The dominant male who is demanding, violent, and self-centered is not considered attractive to most women, whereas the dominant male who is assertive and confident is considered attractive."

-Scott Barry Kaufman ("The Myth of the Alpha Male")

(DOWNLOAD your SPECIAL BONUS "*The High Value Man Checklist*" at www.artofverbalwar.com/thvmbonus)

The Typical Beta "Male"

Now, I wrote in the previous chapter that "alpha males" are not generally thought of very highly or respected by other people.

However, nobody thinks very highly of beta males either, but for different reasons. Why?

Let me tell you a story:

It was my first year in law school. I started dating this girl who was one of the cutest girls in my entire first year law school class of around five hundred people, and a girl that I knew a lot of other guys also liked and wanted to date.

When we started to see each other, I thought I was "The Man".

The relationship started off fine, but as the months went by, I started feeling like she was gaining the "upper hand" in the relationship. Now, other than this gnawing feeling that she was gaining this upper hand, things were still generally fine.

Or, so I thought.

There were some uncomfortable moments during this time, such as two different nights when she got really drunk and then started drunk dialing her ex-boyfriends. One time, one of them even showed up at her place while I was there.

After these incidents, I didn't say anything or change my behavior toward her. I just let it pass.

Or, there were a few other moments, where I'd see her flirting with other guys at school such as at the cafeteria on campus.

I didn't say anything or change my behavior toward her. I just let it pass.

Yet, without any obvious signs of conflict, I still thought things were fine.

At the nine month mark in our relationship, we went to a shopping mall near Stanford University, in Palo Alto, California, to do some shopping.

We got hungry after walking around for a couple hours, so we went to McDonald's to have a snack.

I ordered a Big Mac and fries. As I was enjoying my fries, out of the blue she turned and said to me in a very serious tone "I REALLY HATE how you eat three fries at once."

Stunned and somewhat bewildered by her comment, I muttered "Oh sorry" and complied with what I thought was a demand and ate my remaining french fries, one by one, instead of stuffing them into my mouth by the handful as I was doing.

I still thought things were fine.

However, a week after what I now call "The French Fry Fiasco", she broke up with me. I was crushed.

Being beta is defined largely by over-agreeableness, reactiveness, and being overly accommodating to others, especially women (like I was in this too silly to be true story).

At the time she broke up with me and for the months that ensued, I blamed her. I thought she was a crazy bitch.

I should have looked at myself instead. Instead of handling my girlfriend and the relationship skillfully, I spent over nine months in a relationship letting her step all over me repeatedly. I became less and less attractive to her over time, due to these "beta male" behaviors.

Every opportunity I had to demonstrate attractive male behavior, I failed and instead showed her how much of a "beta male" I was.

In retrospect, reacting to and accommodating my ex-girlfriend's request to eat my fries one-by-one was clear beta male behavior. I showed zero backbone in the face of such a ridiculous statement.

I'm sure this was not the only time I acted like a beta male in this relationship. It was just the most ridiculous and most memorable (in a very bad way) incident.

Once a woman has you marked as a beta male, whether consciously or subconsciously, it is already over for you. She may not break up with you there and then, but it is just a matter of time before she loses complete attraction and respect for you and then wants nothing to do with you.

Now, that is one iteration of beta male behavior, and a rather absurd one at that.

Let's talk a bit more about what a typical "beta male" looks like.

In addition to over-agreeableness, over-reactiveness, and being over-accommodating to others, these are also some traits (among many) that "beta males" exhibit: emotional instability and neuroticism, indecisiveness, oversensitivity, neediness, permission-seeking, non-assertive, lacking self-respect, and spineless.

This is just a short list of behaviors that handcuff beta males from achieving success in life, in their career, with women, and socially.

But, the most defining trait of a beta male is that within a relationship or in relation to group dynamics, a beta male usually is powerless and at the whim of others. A positively masculine man should never be powerless, so beta males are truly out of touch with the masculine side of themselves.

Before we close this chapter, let me share with you one other "beta male" story that's less absurd than my own "beta male" story, but is one that is more common than you may think.

This is the story of "The Blind Beta", or what I like to call him.

There's a guy I know whose lack of self-respect is so non-existent that his girlfriend will flirt with other guys in front of him, including sitting on their laps and resting her head on their chests at parties.

It's not as if this behavior is done in private.

All of this is all done right in front of The Blind Beta, in his face.

Yet, he never calls her out on this behavior. And, she does this repeatedly at parties, week after week, month after month. And again and again, his response (or shall I say, lack of response) is the same. Rinse and repeat.

Behind closed doors, but not the doors to his or her place, she sleeps with and has wild sex with countless other men, including being bent over desks and giving blowjobs to male coworkers while in the office.

The Blind Beta may not know about the guys she sleeps with, but not once has he considered calling her out on her poor behavior that he actually witnesses in person, or considered breaking up with her.

In fact, not too long ago, our poor protagonist actually proposed to her. She turned him down saying she didn't like his proposal. To this day, he continues to be with her, and it is rumored he will soon attempt another proposal again.

I don't presume to know where their "relationship" is heading, but I do know that her sleeping around continues to this day, and will continue as long as they are "together".

I guarantee you that The Blind Beta's lack of self-respect is the reason why all of this started in the first place, and it is the reason his girlfriend's poor behavior continues to escalate.

This is "beta" male behavior at its worst (and saddest), and unfortunately, it is more common than you think. I've got too many similar stories that I know about to tell.

And, while we may all feel bad for the poor saps going through these terrible things, this type of "masculinity", believe it or not, is actually a type of "negative" masculinity.

It may not be the bullying and hyper-aggressive behavior of the "alpha male", but it is negative because the beta male's lack of leadership and assertiveness is what causes these damaged and highly dysfunctional relationships to occur. By failing to assert himself over and over again, the Blind Beta robs his woman of an authentic, powerful leader of the relationship.

Those are just two beta male stories for you, but just know that there is no end to these types of stories.

My hope is that you will never have to be the main actor in one of them.

(DOWNLOAD your SPECIAL BONUS *The High Value Man Checklist* at www.artofverbalwar.com/thvmbonus)

Enter the High Value Man

"True masculine power is full-blooded power-with-power that strengthens both our autonomy and our togetherness, power that is both hard and soft...Such power, whatever its intensity, does not abuse, and protects what needs protecting. It brings out the very best in a man, backing him in taking needed stands, without forgetting his heart. True masculine power is not out to prove anything, but simply to support the living of a deeper life, a life of authenticity, care, passion, integrity, love, and wakefulness."

-Robert Augustus Masters ("To Be A Man")

If you asked a hundred random people what they think the term "high value man" means, most of them probably have certain preconceptions about what this term means.

People most likely misunderstand what a "high value man" is.

It doesn't mean what you think it means.

Being a "high value man" is not about being a super-outgoing, loud, alpha male who drives exotic sports cars, goes clubbing, works in a high-powered, fast-paced job, and bangs lots of women, i.e. a so-called "alpha male".

Having "high value" does not mean making a lot of money, being really good looking, having big muscles, having a ton of friends, and "having it all" generally.

Don't get me wrong. None of these are inherently bad things, but they are all **external** things.

A lot of people believe obtaining and having these external things confer some sort of "status" upon you. And, because alpha males have these things, they therefore have "high status". And, as a counterpoint, beta males have low status.

But this is not true. Yes, a lot of alpha males may have these things, but many do not. And even some beta males may have at least some of these external trappings too.

At the end of the day however, as you will learn later in this book, these **externally facing** things really don't matter.

I want you to stop thinking about gaining "status", but start thinking more about "having value".

Discard the notion that you need to have high "status", discard the old alpha/beta dichotomy, and start thinking about the concept of becoming a "high value man" instead.

A "high value man" is primarily focused on what's internal and inside of him, and not what's external. "Internal" meaning his own internal mental makeup.

Everywhere I look these days, I see men struggling with their identity, struggling to become "men".

In doing so, I see men worrying excessively about their clothes, their haircuts, and whether they will look "good enough".

I see men going religiously to the gym trying to deadlift 500 pounds and talking about it as if once they achieve that goal, they will miraculously become "men". And once they do that, girls will not be able to resist them, or so they believe.

I also see men with no purpose in life other than enjoying themselves, drinking craft beer and seeking out the next hipster pleasure like cronuts, cruffins, and other croissant-related culinary monstrosities.

The funny thing is, becoming a "high value man" can actually help you obtain a lot of the external things I described earlier if these things are important to you. But, being a "high value man" ultimately has nothing to do with these external things.

"High value" is not the value that others place upon you by virtue of the external things you have, although when you are "high value", others will naturally start to place higher value on you.

For the purposes of this book, when we refer to "high value", it is the value you place on yourself.

From here on out, you are going to place a "high value" on yourself.

Becoming a "high value man" is about becoming a man who truly and authentically values himself and his opinion, believes in himself, and takes his own counsel.

And, not in an arrogant or egotistical way like an "alpha male". In seeking to be "high value", you are not going to impose that high value onto others like what an alpha male does in seeking to dominate others.

The way you place value on yourself will be different from that of an alpha male.

And when you place a high value on yourself, the way you naturally behave is also very different from that of a beta male.

"...dominance and prestige represent very different ways of attaining and maintaining status. But it's also worth once again reiterating the overlap: qualities like strength, leadership, kindness, and morality can exist in the same person; strict categories of "alpha" and "beta" truly set up a false dichotomy that obscures what a man is capable of becoming."

-Scott Barry Kaufman ("The Myth of the Alpha Male")

Here are some examples of how a high value man behaves:

A high value man is assertive, but not aggressive or threatening like an "alpha male", and not passive like a beta male.

A high value man cuts to the chase, but in doing so does not abuse others like an "alpha male", and does not avoid conflict like a beta male.

A high value man deflects attempts to manipulate, but does not seek to manipulate or intimidate others himself like an "alpha male", and does not allow himself to be the victim of manipulation like a beta male.

A high value man makes requests of others, but does not violate the boundaries of others like an "alpha male", and does not get ordered around like a beta male.

A high value man cuts through bullshit, but does not bullshit others like an "alpha male", and does not allow himself to get bullshitted like a beta male.

A high value man is comfortable showing respect to others, but does not belittle or put down others like an "alpha male", and does not tolerate disrespect or give undue or undeserved respect to others like a beta male.

To put this all less abstractly, let's go back to the infamous "French Fry Fiasco" of the last chapter.

Upon hearing my ex-girlfriend's declaration that she hated how I ate three french fries at a time, an "alpha male" would blow up at her, go on a rant about how much of a crazy bitch she was, maybe dump his remaining french fries over her head, and then storm out McDonalds in a fit of rage while knocking a couple people out of the way.

A "beta male" would apologize, and then dutifully and awkwardly eat his french fries one at a time in order to appease her. Later that night, he'll still feel awkward about the whole situation, replay it in his mind over and over, and then go to bed without having sex with his girlfriend. The next morning, he'll still feel awkward and avoid her.

Now, this is how a high value man would deal with this situation:

Instead of complying with my ex-girlfriend's "request", I could have handled her by being amused, laughing, and then saying *"Baby...if you didn't know, this is how I LIKE to eat my french fries...three at a time. If my mouth were big enough, I would eat eight at a time. I know you really want to get back to shopping as soon as possible, so let me eat these fries a little quicker"* and then go right back to eating three at a time, as if nothing had happened.

A high value man coolly rolls with the punches that life (or in this case, women) throws at him because he is not arrogant or domineering (like the alpha male) and his self-esteem is solid and grounded (unlike the beta male). His masculine power is always appropriately wielded (unlike the alpha male who always errs on the side of "dominance") and maintained (unlike the beta male who always gives it away).

In this example, a high value man would reframe the situation from her complaining how he eats too many fries at once to how he's not eating enough fries at once and keeping her from shopping. He flips the entire scenario right back onto her.

Now, that's how a high value man takes control of any situation.

You too can learn how to take control of any situation in life, and not how an alpha male takes "control" or how a beta male lets circumstances control him, but in a "high value" way.

Let me close this chapter with a quote from the famous tome "The Way of the Superior Man" by David Deida as we get closer to fully fleshing out the "high value man":

> *"This newly evolving man is not a scared bully, posturing like some King Kong in charge of the universe. Nor is he a new age wimp, all spineless, smiley, and starry-eyed. He has embraced both his inner masculine and feminine, and he no longer holds onto either of them...It is time to evolve beyond the macho jerk ideal, all spine and no heart. It is also time to evolve beyond the sensitive and caring wimp ideal, all heart and no spine. Heart and spine must be united in a single man..."*

High Value Men Focus On The Internal, Not The External

"When you submit in spirit to aggressors or to an unjust and impossible situation, you do not buy yourself any real peace. You encourage people to go further, to take more from you, to use you for their own purposes. They sense your lack of self-respect and they feel justified in mistreating you."

-Robert Greene ("The 50th Law")

Before we get into why becoming a high value man is ideal in the next chapter, I want to tell you another story, this time involving a friend of mine.

I want to emphasize that unlike the "alpha male", whose traits are mostly externally directed, becoming a high value male is completely an internal endeavor.

Becoming a high value man is about cultivating a rock-solid, grounded internal mental makeup.

After spending a number of years in the consulting industry struggling with low pay and crushing hours, this friend of mine was in his mid/late 20s and got a new job where he was suddenly paid quite bit more money than he was getting paid previously.

Feeling flush with money, he spent months deliberating what new car he was going to purchase. Putting aside why it took him so many months making a fairly simple decision, he alternated between a grand vision of a BMW 3 Series and a sobering reality of a practical Toyota Prius, and at the end of this protracted deliberation period, he decided on the BMW because he felt that girls would like that more and he would have more "status" if he drove a BMW.

So, he purchased the BMW.

He was excited for the next few months.

He kept the car nice and clean.

He didn't get any more dates, as he thought he would before he bought his car.

But, maybe it gave him more status?

One day he came into work and found a stack of new, unassembled bankers boxes with a Post-It-Note addressed to him with his boss's initials on it:

"Please put these boxes together and pack the books in my office so they can be taken to storage. -C".

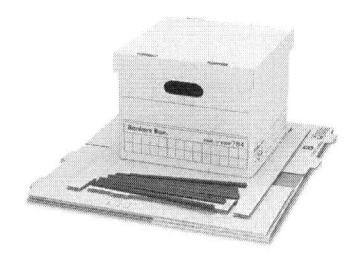

Now, he had a manager title and clearly, putting together boxes and packing things were not in his job description. His boss was known to be quite incompetent with these types of things and probably did not know how to put the bankers boxes together, and that is why she ended up asking him to do it for her.

She was also known to impose on other people and have them do things that were out of their job duties. Some people pushed back on her successfully.

He texted me to tell me about this "request" by his boss. Wait…tell me is the wrong word. Complain is more like it.

I told him that instead of complying with her unreasonable request, he should talk to his boss and tell her that he'd be happy to teach her how to put the boxes together if she didn't know how to.

I felt that would be a good way to deflect her request without making it look like he was being insubordinate. A reasonable solution to his problem, or so I thought.

I texted him at the end of the day to see how talking to his boss went.

He told me that he ended up putting the boxes together himself and not saying anything to her as I had told him to. He said he didn't want to "rock the boat".

This incident became just one of many disrespectful incidents that she subsequently perpetrated on him.

So much for the "status" that his new BMW conferred on him.

My point isn't so much to demean his new BMW, but to illustrate to you that external things have no inherent power to grant you. The BMW, while shiny, impressive, and fast did not make my friend any stronger internally. Unfortunately, in this case, a BMW doesn't make the man. And in any other case, that will also hold true.

Expecting a BMW to enhance a "low value" man's internal mental makeup is about as hopeful as dressing a baby in a Superman costume and expecting the baby to fly, or as ludicrous as putting lipstick and a dress on a pig and expecting it to win a beauty pageant.

In fact, don't ever think that external things have the power to change your internal mental makeup.

And, this internal mental makeup, whether weak or strong, powerful or powerless, radiates from you, and can be felt by others easily.

If your internal mental makeup is weak, people will feel your lack of power, self-esteem, and self-respect. If it is strong, people will also recognize that. In either scenario, they will treat and deal with you accordingly.

The moral of the story: All the power that you want to have in this world is not found outside of you, but inside of you.

Let me finish this chapter with another quote from Robert Greene, which illustrates why becoming high value to the external world is all about first enhancing your internal mental makeup:

"This comes from your attitude - fearless and always prepared to fight. It radiates outward and can be read in your manner without you having to speak a word. By a paradoxical law of human nature, trying to please people less will make them more likely in the long run to respect and treat you better."

-Robert Greene ("The 50th Law")

(DOWNLOAD your SPECIAL BONUS "*The High Value Man Checklist*" at www.artofverbalwar.com/thvmbonus)

The Benefits of Becoming a High Value Man

The reason why many men aspire to become "alpha males" is that alpha males are perceived to be more successful in life, especially in their careers, social lives, and with the opposite sex.

Notice I used the word "perceived".

To the extent alpha males are actually more successful in various arenas in life, which I cannot easily determine, I believe the "high value" male can achieve these same successes without the downsides and anti-social aspects of being an "alpha male" (which we discussed in the chapter "The Typical Alpha Male").

What are some of these benefits?

High value men command respect from others. When you feel like others don't respect you, listen to you, even when you are correct, then it is because others currently perceive you as low value. More accurately, they perceive you as holding yourself with low value, and as a result of this, they give you the same respect that you give yourself.

High value men have more influence than others. If you feel you have a lot to say, but whatever you do say ends up falling on deaf ears, then becoming a high value man will change that.

High value men are admired and looked up to, by other men and also by women. If you feel like people look down on you, it is because you do not have enough value yet in their eyes, and that is because they can feel how much you value yourself through your actions (or lack of action).

High value men's opinions are highly valued. If your opinion seems to be ignored by others regularly, it is likely others are perceiving you as low value, and as a result, that is how they perceive your opinions.

You will find that things in life "just come easier" for high value men. If you feel like people usually stand in your way and are obstacles for you to achieve what you want, or they make life difficult for you, then it is because you are not yet a high value man, and others can sense that and treat you accordingly.

Not only do things come easier as a high value man, but people will treat you better.

I used to think that when people disrespected me or women didn't treat me the way I wanted them to, that the world was unfair, or that a particular man or woman were assholes, jerks, or bitches.

That is just not true.

People treat you exactly the way you **show them you should be treated**. It is **not** fairness that dictates how you are treated in life by other people.

The world becomes a lot more "fair" when you become a high value man, I promise.

Sounds good, doesn't it? So, without further ado, let's start learning how you can become a "high value man".

(DOWNLOAD your SPECIAL BONUS "*The High Value Man Checklist*" at www.artofverbalwar.com/thvmbonus)

CHAPTER TWO

The Six-Part Plan For Becoming a High Value Man

Becoming High Value

Part 1: Eliminating Negative Behaviors

"So knowledge grows by subtraction much more than by addition."

–Nassim Nicholas Taleb ("Antifragile")

Becoming a high value man is a process, not an overnight transformation, and it has six parts.

Let me begin this first part by declaring that, as in the quote above in the famous book "Antifragile", becoming a high value man is actually largely about shedding negative or value lowering behaviors (i.e. "subtraction") and less about adopting any specific behaviors or character traits like you see so-called "alpha males" doing (very unnaturally I must add) (i.e. "addition").

You start the journey of becoming a high value man from wherever you are by eliminating these negative or value lowering behaviors that signal to others that you feel that you are "not good enough".

When you make substantial progress in this very large step, others will come to see you as a completely different person.

Becoming high value is primarily about handling the interpersonal dynamics you experience with other men and women. If you can succeed in this mission by maintaining and demonstrating your value, you will become and be perceived as a "high value man".

What I'm saying is that negative or value lowering behaviors are so detrimental that when you exhibit these behaviors, other people will not be able to see or appreciate the better and more interesting sides of you that are obscured by these negative behaviors.

Here are the TWELVE negative, "value" destroying behaviors that must be eradicated from your default behavioral systems with the vengeance of a nuclear weapon.

Starting today, let's declare war on these twelve behaviors:

1. Seeking Permission To Act / Validation For Acts Taken

High value men do not seek the permission of other people to do the things they want to do.

It is not that seeking advice is always a bad thing; in fact, it can be a wise thing to do, but the truth is, when you are a grown ass man, you do not need another person's permission to do anything.

So, if you are a grown ass man, then start acting like one.

People who seek permission tend to hesitate unnecessarily and miss out on great opportunities in life. Give yourself permission to do the things you want to do or have to do.

When you are a child, you must seek your parents' permission to do things. But when you're a grown ass man asking people for permission or validation for your decisions, you should not wonder why people treat you as a low value man.

In addition to permission to act, you do not need anybody else's validation for the choices you make for yourself. Once you have acted, resist the need to get confirmation from others about the things you have done. What good is their confirmation anyways? What is done is already done.

You don't need a pat on the back from others and the truth is, as an adult, when you seek such validation from others, any "pat on the back" you may receive lacks authenticity, as others will undoubtedly perceive your need for validation in a negative way. Don't fool yourself into thinking that false accolades mean anything.

2. Overly Seeking Other People's Opinions

Similarly, a grown ass man should know what he wants in life, what he wants to do, and provides counsel to himself.

Ultimately, nobody else knows you the way you do, so other people's opinions about your life and your plans are irrelevant.

Whenever he had a major decision to make, one of my best buddies from college would always call his sister or mother to seek their advice, and usually both. He would not do anything without their opinion and their approval.

As a grown adult and now as a married man, he now has a new "boss" in addition to his two existing ones, and it is very apparent. His indecisiveness persists more than ever, and due to the fact that he has leaned on others his entire life, all of his life choices are now dictated by others, and he has lost his ability to make decisions for himself.

Even asking him to make a simple decision such as where he would like to eat is an excruciating experience to witness.

He has lost his ability to be entirely self-sufficient, and it is a sad thing to witness.

Learn to listen to your own voice and shut out voices that cannot speak for you. Stop being indecisive, and get comfortable making your own decisions and sticking to them.

If people try to make decisions for you or impose their opinions on you, cut them off (in a constructive, high value way of course).

3. Overly Apologetic

Back in high school, I used to be friends (the key phrase here is "used to be") with this guy who for whatever reason, would apologize about a hundred times a day to anybody who would listen to his self-flagellation.

He was so "apologetic" that if he could, I bet he would apologize for living and breathing. I would bet money that he has actually done that many times, but I just wasn't there to witness it.

And, I hope I never have to.

Being overly apologetic is an extremely low value behavior. Here's a beta/low value male ON HIS KNEES in Hong Kong apologizing to his girlfriend for whatever transgressions he committed. All the while she's slapping him in front of other people in public.

You don't want to ever be this guy. This is as low value as a man can get. Even if you've done something wrong, nothing warrants you getting on your knees and letting someone slap your face.

There is nothing wrong with apologizing for things you have done truly wrong to another person, but never apologize for having an opinion, doing things that benefit yourself, and for things you have no control over. When apologizing, you don't have to be submissive and self-flagellating.

And, most of all don't apologize for living!

4. Blindly Accepting Others' Opinions

In life, people will always try to impose their viewpoints and opinions on you. It is okay to listen to and respect other people's opinions, but don't accept them blindly or even easily.

When you set a pattern of behavior of deferring to other people's opinions, but then you suddenly have an opinion on something, people will react to you having an opinion violently. So, the best thing to do is to never set an expectation that you will defer to the people in your life.

Low value men have a very strong tendency towards following and deferring to others. You don't have to always dominate people and be selfish, but you do need to check this tendency so that people will learn to respect you.

Instead, make sure you question others on their viewpoints and opinions. Ask for proof and evidence, and don't get caught up in the logical fallacies and dirty rhetorical tricks people use. If you find yourself having problems with this, check out my book "The New Art of Being Right".

Accepting other people's opinions too easily is a low value behavior. If you are a high value man, you should not feel bad about questioning viewpoints and opinions that you do not agree with since you have your own valid opinions.

It is just a switch in perspective that's required here. Instead of always letting others dictate their opinions to you, you should start learning how to bring your opinions to others and letting them respond and justify their opinions to you.

Now, don't go too crazy with this, but be at least a bit of a challenge to other people when appropriate. When they understand that they cannot steamroll you with their viewpoints and opinions, they will develop more respect for you in the long run.

5. Excessive Humility (or Excessive Bragging)

When you have done something worthy of praise or of high achievement, don't downplay your achievement. You don't need to be excessively humble about what you have accomplished.

Some people take humility to such an extent that they even engage in self-deprecation. Don't self-deprecate! Never put yourself down in front of other people!

On the flip side, you also do not want to excessively trumpet your accomplishments either (like The Black Belt Bully I referred to in a previous chapter). Live by the saying "Act like you've been there before".

When you brag about your accomplishments, don't think you are fooling anybody. Nobody is impressed by a braggart, and demonstrating a need to brag is an obvious poker tell for insecurity.

Being on either end of the spectrum is low value behavior.

6. Avoiding Conflict / Non-Assertive / Easily Intimidated

Don't get into unnecessary conflicts, but when it comes to unwarranted disrespect, don't back down and be assertive with your own needs and desires.

Remember the banker's box incident in a previous chapter? Remember how failing to confront his boss caused my friend to endure numerous subsequent disrespects? You don't want to always be fighting, but you cannot avoid conflict when it is necessary to protect yourself from a potential pattern of disrespect.

You have to be willing to (or at least show that you are willing to) confront people when they have wronged or disrespected you, especially those people who you must interact with frequently.

The goal is to give people the impression that you are not someone who can be trifled with or intimidated. A high value man is not someone who can be trifled with. Let them know they will have to pay a heavy price if they decide to trifle with you.

7. High Neuroticism

Numerous studies of status in social groups (fraternity, sorority, and dormitory) and peer ratings of status determined that "high neuroticism" predicted lower status in men.

Neuroticism is basically a variety of negative emotions such as anxiety, envy, and jealousy. Or, in simpler terms "emotional instability".

If you show anxiety, envy, or jealousy, that is inherently a low value behavior. If you are truly a high value man, these emotions will not come to you naturally because your cup is always overflowing.

Anxiety, envy, and jealousy come naturally to people who question their own value, whether consciously or subconsciously. When they see others doing well, that also causes them to question themselves.

However, a high value man is not bothered or triggered by the successes of others, and that is because he holds himself in high regard.

Learning how to manage these negative emotions is beyond the scope of this book, but if you suffer from any of these emotions consistently and to a great extent, I encourage you to seek help from professionals who can help you work on these issues. It doesn't have to be this way!

8. Overly Reactive / Over-Sensitive

By "overly reactive", I mean too sensitive to other people's actions or omissions against you.

When people commit small slights against you, you take it as if they have slept with your mother, and you lash out as if this were the case. Sometimes, people may not even realize they have slighted you, yet you immediately jump to this conclusion.

When people don't do the things you may expect from them (but which they may not even know you expect), you harbor resentment and anger, and act passive-aggressively.

Or, you just let other people throw you off easily or get under your skin, and worse yet, you show how easily triggered you are and everybody can read this characteristic of yours like a book.

All these things are low value behavior.

"Alpha males" actually have a problem with this one. Don't fool yourself into thinking you are acting as a truly masculine man would, or in a manner that shows you cannot be trifled with. Your short fuse is a clear poker tell to the world you do not have your house under control, i.e. your internal mental makeup.

High value men do not let small slights or incidents bother them, and these small slights bounce off of them as if they were made of Teflon.

As of the writing of this book, there was a recent "road rage" incident where a famous athlete was shot to death by another driver.

The facts of the case are not yet completely clear, but I suspect this road rage incident has to do with egos being bruised too easily, tempers being unreasonably short, and small slights being taken as large insults.

The lesson to take from this is that having celebrity status and big muscles are external things, and these things as usual, do not necessarily reflect one's internal mental makeup.

9. Uncomfortable Around People

Low value men wear their discomfort in social or work situations on their sleeves. You can tell in their body language, eye contact, vocal tonality, and all other sub-communications they give off.

Often times, they view other men as threats because of the higher value they perceive these other men to be. This causes them to act uncomfortably.

Around women, they act awkwardly and subconsciously elevate women to a pedestal. Obviously, this is highly unattractive to the opposite sex.

Studies have actually shown lower "status" men in groups of people tend to have higher cortisol hormone levels ("cortisol" being the 'stress hormone' that is released when humans are stressed).

Ability to handle the "stress" of social interactions has been shown in studies to be correlated to status hierarchies through the studies of how long people maintain eye contact. When people meet each other, there is a "nonverbal dominance contest" that occurs (usually subconsciously). What I mean by this is that the person who breaks eye contact first usually ends up as the "lower status" person in that relationship.

Also correlated with higher status in these studies are lowered brows and a non-smiling mouth. The theory is that "status" within groups of people is merely a reflection of the underlying "dominance dynamic" based on the differences in "outstressing" ability of the people involve.

Now, don't go crazy with trying to implement these findings in these studies, but the takeaway here that you must develop what's called your "outstressing ability" muscle over time.

What do I mean by that? You must learn how to become more comfortable and relaxed around other men and women. You must learn to be relaxed in the face of social tension. Got it?

10. Try Hard

This isn't completely intuitive or easy to develop, but the most high value men always appear to be effortless, relaxed, and easy going.

Low value men on the other hand, always come off as tense, trying too hard to fit in, trying too hard to be perfect, and as a result, everything they do appears to others to require tremendous effort.

This is a simple precept, and it sounds zen-like, but not easy to actually implement:

The harder you try, the lower value you appear.

The less you try, the higher value you appear.

You may want to become more high value, but most of the time, if you try too hard, you will actually come off low value. Getting rid of this behavior will require you to let go, and to care less about your own value and the value of others.

Only when you do that will your find your value growing.

11. Prioritize Being Liked

This is one of the main ways low value men undermine themselves when it comes to interpersonal relationships.

Low value men trade away their own value and being true to themselves in order to be liked. Wanting to be liked and acting in accordance with this desire leads men to lose their "power" with respect to interpersonal relationships.

For example, when it comes to women, a low value man would rather that she "like" him and that he feels comfortable, than risking building any tension in an interaction with a woman.

Therefore, he plays it safe and fails to develop any sexual tension with a woman. He would rather not "lose" her than play to win. This is just like the saying "winning the battle, but losing the war". By prioritizing his need to be liked, he has already lost her.

With other men, a low value man doesn't stand up for his own opinions. He would rather agree with other people's opinions rather than challenge them when he does not agree. His main priority is to fit in, rather than express himself freely.

With all people, low value men aim for "perfection". They try not to make any mistakes, hide their perceived "flaws", and become completely bland, uninteresting, and low value, all in the effort to be "liked".

"Humans connect with humans. Hiding one's humanity and trying to project an image of perfection makes a person vague, slippery, lifeless, and uninteresting."

-Dr. Robert Glover ("No More Mr. Nice Guy")

All of these things contribute to a complete lack of self-respect. If you seek to be "liked", you will undoubtedly encounter numerous situations in life where people disrespect you, whether in a major or minor way. Regardless, if you do not stand up for yourself and instead take the disrespect in order to avoid tension or to continue to be "liked", you will take the hit, i.e. lower your own value in other people's eyes.

By standing up for yourself, you maintain and grow your value over time. People will be forced to respect you or stop dealing with you altogether. I say THAT is a good thing.

12. Neediness

Neediness can come in many forms, such as needing other people's permission or validation (as we discussed earlier), opinions (also as we discussed earlier), their time, or their positive emotions.

Low value men are extremely needy because they subconsciously do not feel they have any value of their own and so they (whether consciously or subconsciously) always look to leech other people's value, which may include their validation, their opinions, their time, and their positive emotions.

Ultimately, nobody respects a needy man, so if you think you exhibit this behavior, you must eradicate it since it is truly one of the most attraction killing and value lowering behaviors of them all.

Let me close this chapter by saying once and for all: Neediness MUST no longer be one of your guiding emotions in interpersonal relations.

(DOWNLOAD your SPECIAL BONUS *"The High Value Man Checklist"* at www.artofverbalwar.com/thvmbonus)

Becoming High Value

Part 2: Two Guiding Principles

"People who have a solid sense of their own value and who feel secure about themselves have the capacity to look at the world with greater objectivity. They can be more considerate and thoughtful because they can get outside of themselves. People with a strong ego set up boundaries — their sense of pride will not allow them to accept manipulative or hurtful behavior. We generally like to be around such types. Their confidence and strength is contagious. To have such a strong ego should be an ideal for all of us."

-Robert Greene ("The 50th Law")

Many of you are probably already familiar with "Pareto's Principle."

If you are not familiar with this concept, "Pareto's Principle" is a principle, named after economist Vilfredo Pareto, that specifies an unequal relationship between inputs and outputs.

Unequal relationship between inputs and outputs?! What does that mean?

Pareto's Principle states that 20% of invested input (i.e. you effort) is responsible for 80% of the results ("outputs") obtained, and this principle applies very closely to the journey towards becoming a high value man.

Becoming AWARE of the behaviors I wrote about in the previous chapter will take you 80% of the way towards becoming a high value man even though it will only take you 20% of the effort to develop this awareness.

A lot of people are simply just unaware of their behavior. Not only are they unaware, they do not know that such behaviors are "low value". Society has brainwashed many men into thinking that certain behaviors are attractive, when they absolutely are not.

As for the other 20% of the way, a lot more effort will be required, i.e. 80% of the effort, which is the step beyond awareness of these low value behaviors, but replacing these low value behaviors which have become part of your default behavior with new, high-value behaviors.

In this chapter, you are going to learn some new behaviors and character traits that you need to start adopting to become part of your internal operating system.

In the previous chapter, I wrote about the twelve negative behaviors one must eliminate. That was an incredibly imposing list! It's a very tall order to start becoming the opposite of any of these traits, even just one.

So, I want to make doing that as simple as possible for you. I believe that becoming "high value" simply boils down to living your life guided by the following two crucial guiding principles:

#1

At the heart of becoming high value is this central guiding principle. Simply, you must start acting as a high value man would act in every moment of your life, like the type of man described by Robert Greene in the above quote.

So, in any given moment, ask yourself this question:

Am I behaving as a person who values himself, his time, and his opinions, would in this moment?

Whenever you do not know how to deal with any given situation, ask yourself this question and act according to how you think a man who values himself, his time, and his opinions would act.

One thing is for sure, a high value man will put his needs first, then consider what others may need. A beta male would always put others people's needs first, and an alpha male would consider his needs first and solely.

Think of "high value" behavior as a muscle of yours that needs to be developed, just like any other muscle on your body, like your biceps or quads.

At first, in the physical realm, lifting any weight (even a small dumbbell) will be difficult. But if you keep at it, you will undoubtedly be able to lift more and more over time. After some time, you will suddenly be putting up some impressive numbers. This is how every Olympic strongman starts.

Now instead, we are going to make "high value behavior" the "muscle" you are going to develop, but this time in the mental realm.

Of course, this will not be easy at first, as you may not necessarily know how a high value man should act.

At first, you will find yourself reverting to your default, low-value behaviors frequently. That is okay. Don't beat yourself up and keep working at it.

This is a transformation of the mind more than anything, and that will take time. If you put in an honest effort over time, I promise your default low value behaviors will start getting replaced by high value behaviors.

In a year's time, you will find your reaction to circumstances in life, especially negative ones, will be completely different than they are now.

#2

The other crucial concept which I actually brought up earlier in another chapter which I want to expound on further is that becoming "high value" is completely about developing internal mental fortitude as opposed to carrying out "external displays of dominance" and acquiring the trappings of "success" of an "alpha male".

By treating yourself with high value, it will show in your actions especially in relation to how you interact with people and how people interact with you.

Here is the crucial mantra you should repeat to yourself over and over:

What's most important is *WHO I am, and not WHAT I do or have.*

You are now someone who will put a "high price tag" on yourself, which signals to others how you treat yourself and how they should treat you.

When you first meet another person, people will pick up on what value you place on yourself. They will have no idea what kind of person you are, so your job is to demonstrate the type of person you are (and how much you value yourself) in how you respond to them and what they notice about you (the "WHO"). You will no longer focus on meaningless external things (the "WHAT") like the things I wrote about in previous chapters.

And, if you are dealing with people you already know, you need to change your behavior towards them. At first, it may be difficult as they already have expectations about your behavior, but you must not be deterred by their initial inability to comprehend your changed behavior and their initial resistance to change.

For some of you, let me warn you in advance that this process of changing expectations may cause you to lose friends who cannot adjust to your new behavior.

Your low value self may bemoan losing friends and prioritize being liked over being respected, but your high value self will not allow you to do that.

For every person it is a personal choice whether to accept the results of putting higher value on themselves. I cannot speak for you, but for me, I would rather be respected than merely liked or tolerated.

I leave this decision to you, but for me, it is an easy choice.

Becoming High Value

Part 3: Value Enhancers

In this chapter, we are going to talk about some other behaviors you can start implementing which have been shown through numerous research studies to help increase your "perceived value" to others.

I know that I wrote that being "high value" is mostly developing an internal mental makeup that is "high value", and that is absolutely true.

However, there are a number of things that are scientifically shown to enhance the "high value" you display to the world, which I will call "value enhancers". Developing a high value mental makeup is a process, but learning and implementing these "value enhancers" will help you ignite this process and see some earlier returns.

These value enhancers are like the MSG added to Chinese food in order to enhance its flavor, except without the gastrointestinal distress, thirstiness, and general ickiness.

Sure, you can ask for "no MSG" when you order the food, and it will still be delicious, but there is no doubt that the MSG flavored food is just a little more tasty.

In my opinion, these value enhancers are things you can start using immediately while you build your internal mental makeup into that of a "high value man", so sprinkle these here and there, and I promise you will also see people start to perceive you differently and treat you better.

1. Speech and Voice Tonality

According to studies, individuals' speech can serve as a cue to both their external status in society and their relative status in an interaction.

In particular, people with higher external status are likely to talk more than individuals with lower external status.

And, not only do these people spend more time talking, but surprisingly, they also spend more time pausing.

Pausing is a sign of "controlling the conversation". In these studies, lower status people spent less time in pauses, so it is hypothesized that silence, i.e. making people wait on your words, as well as talking, are indicators of status.

In other words, in an interaction, the amount of talking you do is correlated with your status within an interaction.

The long and short of it is, don't be a wallflower. Make sure you do your fair share of talking and don't be afraid to "hold the floor" by pausing when you do speak.

In addition, not only do you want to speak as much as you reasonably can, you also want to speak authoritatively and with certainty. As I wrote in my book "Vocal Superstar: How To Develop a High Status Voice", you should work on your "vocal tone consistency" in order to appear more "high status".

Back when I was in high school and college and undoubtedly "low value", I remember that every time I talked to a girl, the pitch of my voice would change and go higher and softer.

Embarrassing! But, don't tell me that this has never happened to you!

Studies have shown that in an interaction, people who changed their vocal tonality more to accommodate another person's vocal tonality were lower status compared to the other person.

What this means is you should pay extra attention to ensure that the pitch and tone of your voice do not change when you interact with other people, even those who you perceive to have higher status than you.

2. Extraversion / Perception of Extraversion

One of the most important goals and outcomes of social life is to attain status in the groups to which we belong.

Such face-to-face status is defined by the amount of respect, influence, and prominence each member enjoys in the eyes of the others.

Based on studies of college social groups (such as fraternities, sororities, and dormitories), high extraversion is correlated to elevated status for men and women.

If you're introverted, this doesn't mean you are destined for life as a low value man, but just keep in mind that you need to display some extraverted characteristics at least occasionally.

Just like the studies on speech and vocal tonality showed that the amount of talking predicted status among people in a social interaction, extraversion is also similarly correlated.

Let me reiterate this again: Don't be a wallflower if you want to become a high value man. High energy, extraversion, and expressing yourself to others are necessary from time to time if your goal is to be a high value man.

3. Eye Contact

Studies have shown that higher value people tend to look at other people more while speaking than low value people do. In other words, when speaking, lower value people tend to not make eye contact with the other person.

Other studies have also shown that when you first meet someone, the person that maintains initial eye contact the longest tends to end up the higher perceived value person in that relationship. So, start practicing not breaking eye contact with people when you first meet them until they break eye contact with you first.

Let me simplify this for you: When speaking, try to look at people in the eye more than you do already.

When listening, looking at people in the eye is not as important as when speaking.

To make it even simpler: Under all circumstances, more eye contact is better than less!

4. Body Language

Body language is also a major enhancer of your perceived value, but improving your body language is beyond the scope of this book.

There are a lot of resources available out there on body language, and since this is a very broad and deep topic, I will not be focusing deeply on this value enhancer.

However, here are just a few other body language pointers to help you get started in enhancing your perceived value to others:

1. Always keep your hands away from your face.

2. Don't fidget, including playing with your hands and feet, like tapping your fingers or toes.

3. Avoid checking for reactions whenever you say something.

4. Let yourself be "vulnerable" physically, such as exposing your neck and torso to people.

5. Consistent with the previous tip, don't fold your arms across your chest. Always think "open body language".

6. Always try to move comfortably and gracefully. This also means avoiding making abrupt, herky jerky movements, but instead, move in a slow, measured, and controlled way.

7. Always check your posture. Does it show that you hold yourself in "high value" or does it not? But while you want to always have good posture, don't be too stiff. You want good, but relaxed posture or else you may risk coming off as anxious.

8. When sitting, make sure you spread your arms and legs out wide. Don't go too crazy with this, but take up a comfortable amount space without being rude about it.

I know this is a lot to remember and implement.

Ultimately, what's important to remember from this chapter on value enhancers is what I talked about in a previous chapter, which is:

In order to become more "high value", a major key is to develop your "outstressing" ability muscle over time.

If you think about it, all of these "value enhancers" go towards showing outwardly to other people that you have strong "outstressing" ability.

In order to implement these value enhancers, don't try to implement them all at once. Pick one or two things to work on at any given time and practice them for at least a week or two.

Then, pick another one or two things and practice those for at least a week or two, and keep repeating this process until all these things become second nature.

And, if you ever find that you have trouble implementing all of these value enhancers, just focus on "outstressing", which means focusing primarily on staying as relaxed as possible in interactions with other people. Your mantra should be "remain as comfortable as possible". That is the KEY to appearing high value.

Someone who is comfortable will act in a certain way. Their body language will be relaxed and take up space. You will speak in a measured and assured tone. Your posture will be good. And, you will say whatever you feel like saying, as opposed to editing your words relentlessly. That's what a comfortable person looks and sounds like.

Becoming High Value

Part 4: Aggressiveness vs. Assertiveness

"Taken together, the research suggests that the ideal man...is one who is assertive, confident, easygoing, and sensitive, without being aggressive, demanding, dominant, quiet, shy, or submissive. In other words, a prestigious man, not a dominant man."

-Scott Barry Kaufman

A lot of the previous chapters on becoming high value pre-supposes that you are coming from a place where you are a low value or beta male, which means you are trying to build your assertiveness and self-respect from a point where it is minimal or on the low side.

However, what if you are coming from a place of being a so-called "alpha male" or "bad boy" where your default way of interacting with the world is unchecked aggressiveness, and now you desire to express your masculinity more positively?

Or, maybe you are coming from being "low value" or "beta", and in trying to become more assertive, you are finding it too easy for the pendulum to swing too far the other way?

In my life, every time I've failed to balance the fine line between aggressive and assertiveness, things have not turned out well for me just like my old boss that I talked about in a previous chapter, so it is fundamentally important that a high value male learns how to traverse this fine line.

"Across three studies, the researcher Lauri Jensen-Campbell and colleagues found that it wasn't dominance alone, but rather the interaction of dominance and pro-social behaviors, that women reported were particularly sexually attractive."

*"They found that only one woman out of the 50 undergraduates in their sample actually identified "dominant" as one of the traits she sought in either an ideal date or a romantic partner. For the rest of the dominant adjectives, the two big winners were **confident** (72 percent sought this trait for an ideal date; 74 percent sought this trait for an ideal romantic partner) and **assertive** (48 percent sought this trait for an ideal date; 36 percent sought this trait for an ideal romantic partner). Not one woman wanted a demanding male, and only 12 percent wanted an aggressive person for a date and romantic partner."*

-Scott Barry Kaufman, quoting the above studies

Assertiveness is properly channeled and confident communication that is aware and respectful of other people's needs, position, and authority, whereas aggressiveness is improperly channeled and out-of-control.

In all the challenging circumstances in life you will face, your goal is to express and advocate for your own needs and position while also being aware of and respectful of other people's needs, position, and authority. This is true assertiveness, whereas aggressiveness is expressing your own needs in a manner that either ignores other people's needs, position, and authority, or is destructive of the relationship at hand.

Assertiveness is a way to get your way "charismatically", instead of by intimidation or brute verbal force. You may think that being an "alpha male" is your ticket to success, but know that at the end of the day, according to researchers, it is "dominance **and** pro-social behaviors" that is most attractive.

So, let's go back to The French Fry Fiasco.

The beta male in me at the time took my ex-girlfriend's comment that I was eating too many French fries at once too seriously. Due to the low value I had of myself, I took it as criticism. And worse yet, I did not speak up when this incident happened. I just complied with what I thought was a demand. This is passivity and very unattractive to its core.

An alpha male would also take my ex-girlfriend's comment as criticism. However, instead of passivity, his reaction would be different. The alpha male would get angry, yell at her, and then storm off in a fit of rage. This approach is unchecked aggressiveness.

If given a choice between a beta male and an alpha male, some women would still prefer the alpha male because he stands up for himself. But, most women with a reasonable head on their shoulders and any decent amount of self-respect would also not prefer this archetypal, out-of-control man.

Now, the high value man would handle my ex-girlfriend in a completely different way. The high value man would handle her in an assertive, charismatic, but non-aggressive way.

While the alpha male and the beta male would both take her comment as criticism, the high value man would actually find her comment amusing.

Same comment, different reaction and viewpoint.

Due to his different perspective of her comment, his reaction to it would be entirely different than that of the alpha male or beta male (ironically, they would both view this incident in the same way). The reason he can take her comment in an amusing way is because of the high value he places on his opinion of himself.

A minor comment gets brushed off like a piece of lint. It takes a Herculean effort to throw a high value male off from his path, and a stupid comment from a girl doesn't pass muster.

A low value man can only take her comment in one way, which is as criticism, and that is because his view of himself is already low, so even such a minor comment from a girl he has placed on a pedestal would devastate his self-image.

And paradoxically, the alpha male's view of himself is so grandiose, unrealistic, and based on puffery, that he also perceives her comment as criticism. That's an illustration of the paradox that is the alpha/beta paradigm.

Passiveness is not the answer.

Aggressiveness is also not the answer.

Assertiveness is the key to transcending the alpha/beta paradigm and unlocking the secrets of "positive masculinity".

Becoming High Value

Part 5: The High Value Man In Practice

The previous four parts to "becoming high value" were a bit heavy on theory and principles.

In this next to last part to "becoming high value", let's get more practical about how to conduct yourself as a new "high value man".

I want to make sure that you understand how all these principles work in practice in real-life situations, so that you can rely on these principles in the real-life situations you encounter going forward.

The following are two short excerpts from "No More Mr. Nice Guy" by Dr. Robert Glover about the behavior of an "integrated male", which I feel is great guidance about how a high value man should act in real-life situations:

> *"He is clear, direct, and expressive of his feelings."*

> *"He knows how to set boundaries and is not afraid to work through conflict."*

So, using these guidelines from Dr. Glover and the principles I set forth earlier, here are a few examples of practical situations that men frequently face, and how your typical beta male, alpha male, and high value male would respond:

1. Someone cuts in front of you in line

You are at the grocery store waiting in the checkout line. While you are waiting and looking at your phone, somebody else cuts in front of you.

Beta: Roll your eyes, act generally annoyed, but say nothing.

Alpha: "Hey asshole/bitch, I was here first!"

High Value: "Excuse me, I was here first. I'm going to assume you didn't see me. You can get in right after me."

2. Deciding where to eat

A low value man even has trouble deciding where to eat with friends. Let's say you actually want to eat burgers, but your friends have other ideas.

Beta: Say nothing about wanting burgers. Or, "I'm okay with anything. Really, I am."

Alpha: "I want burgers today, and I want Burger King." (while not acknowledging that other people may also have desires and preferences)

High Value: "Today's feeling like a burger day, but I'll listen to some other ideas and if we don't get burgers this time, let's make sure we get burgers next time."

3. Doing favors

You are at a party and the drinks are in the other room. You and a girl you're talking to are the same distance from the other room. She asks you to go get a drink for her.

Beta: You dutifully get up and go a drink for her from the other room.

Alpha: "You can get it your damn self, I'm not an errand boy."

High Value: "I'm pretty sure you can get a drink yourself, but hey we can take a walk over there, see what's going on, and get the drinks together. Let's go."

4. Saying no

A friend, who is a girl, asks you to help her move over the weekend. You are not sleeping with her, nor do are you in a relationship with her. In fact, you don't even know her that well.

Beta: You help her move, even when you had something else scheduled on the day she planned her move. You were hoping maybe she would pay you back in some way, but of course, she doesn't. You harbor resentment.

Alpha: "Oh hell no. Just find someone on Craigslist or some sucker to help you."

High Value: "I don't help friends move when they're working and making money and they can hire someone to help them move. In fact, I said that to my friend the other day when he asked me help him to move. But hey, I've got a few movers I can recommend to you."

5. Money matters

You lent your friend some money at a group dinner to pay for his meal. He said he would pay you back, but after seeing him four times, he still hasn't paid you back.

Beta: Say nothing, maybe hint at it, but never address the issue directly.

Alpha: "I need my money back, you never make good on your debts. I'm never going to lend you money again!"

High Value: "Hey remember when I lent you money for dinner a few weeks ago? You might have forgotten about it, which is fine this time, but please don't forget to bring money next time I see you."

6. Bad moods

Your girlfriend or wife frequently has bad moods, more often than she should. You think she creates mountains out of molehills.

Beta: Try to "fix" her bad mood, always ask her "What's wrong? What can I do to help?" Offers unsolicited advice and worries more than her.

Alpha: Does not feed her bad mood, gets angry at her for being in a bad mood.

High Value: Does not feed her bad mood. After seeing her pattern of behavior of making mountains out of molehills, the high value male let's her work it out on her own.

7. Asking a girl out

You want to ask a girl to go with you to have lunch or dinner at your favorite restaurant.

Beta: "When are you free? Are you free this weekend? Oh, and do you like sushi?"

Alpha: "I have one hour free on Friday night. I'll squeeze dinner with you in."

High Value: "I feel like going to my favorite sushi restaurant this weekend and I'm free on Saturday. Come with me, I want you to try this amazing dish that's gonna change your life."

8. Late arrival

A girl arrives thirty minutes late to meeting you for drinks. She doesn't apologize for it. This is not the first time she's done this.

Beta: Says nothing about her being late. Then, he continues to act as if nothing has happened.

Alpha: "Why the hell were you late? You're lucky I didn't leave!"

High Value: "This is the second time you've been late to meeting me. Both times I was on time. If we're going to keep seeing each other, please do what you have to do so there won't be a third."

9. Too much texting

This is an unfortunate outgrowth of the modern age of technology.

While meeting you, a girl is constantly distracted by her phone and spends more than half the time texting with someone. She constantly asks you to repeat what you said.

Beta: Act a little annoyed, but says nothing.

Alpha: "Are you setting up another date while we sit here? What the hell!"

High Value: Wait for an opportunity where she asks you to repeat yourself and say "This is the third time I've had to repeat myself. I'll let you finish what you're doing and then we can talk without distractions."

This is just a limited selection of real-life, practice situations that you may face in life, but I hope I have given you enough here to help you start coming up with your solutions to the unpredictable things that life throws at you.

Commentary

Remember that the purpose of responding like a high value man is to ensure that you are maintaining your value with respect to that other person. When you let bad behavior or boundary violations occur without addressing it properly, you are sending a message that the other person can continue to treat you as a low value person.

In all of the above examples, note the following principles being played out skillfully by the high value man in being assertive, but not aggressive or passive:

1. Setting forth your principles, setting them forth explicitly, and making sure it is clear that they should be attributed to you (and not someone or something else), i.e. taking ownership.

2. Avoid histrionics and unnecessary emotion and drama.

3. Stating your complaints in a matter-of-fact way.

4. Calling out bad behavior in a way that leaves the other person a bridge to get in your good graces again.

5. Stating your needs, but also leaving room for some flexibility.

6. Making sure the other person knows you are being forgiving of their bad behavior, but only on the condition that it be rectified.

7. Leading the other person to (or giving them an opportunity to reach) an outcome that you desire.

Becoming High Value

Part 6: Exercises and Tools

Let me close this book with five exercises and/or tools to help you reinforce the principles and concepts you learned in The High Value Man, and to implement them on a daily basis.

Some of these exercises are done daily, some are done on a quarterly basis, and some are more like tools that you use as needed, i.e. when you need.

If you've come this far, please don't skip these exercises as they are an absolutely crucial part of the PROCESS!

Exercise 1: Setting Boundaries (every quarter)

Review the areas in your relationships with people (friends, family, work colleagues) where you may currently be avoiding setting appropriate boundaries or avoiding conflict.

We all have them, including myself, no matter where we are in life. And, that's okay, but we can't let them fester.

So, write them down right now.

Here are some specific areas that you should look at:

1. You are tolerating behavior that is toxic, manipulative, or problematic.

2. Avoiding dealing with important situations because you are afraid it might cause conflict.

3. Failing to ask for things you want or need in a relationship.

4. You are resentful towards another person.

Now that you have written them down, think about how you will address these situations using the principles you learned in this book.

Write out how you will do that for each specific situation, and then in the next 1-3 months, start addressing each of these situations one by one.

Exercise 2: Setting Personal Rules, Standards & Principles (every quarter)

In this exercise, I want you to write out rules, standards, and principles that you will live by going forward. Let's call these your "high value rules".

To get you started, here are some ideas:

1. If it frightens me, I will do it.

2. I will put myself first, then consider other people's interests, opinions, and concerns.

3. I will ask for what I want in a constructive way.

4. I will learn to say no when I feel that the answer is no.

Write down at least ten rules and make it the lock screen image on your phone, or put it up on the wall in your office, or anywhere you will be able to see it often.

The more specific you make each of these rules, the better. So, for example "I will stand up for myself" is not very useful, but "Every time my girlfriend criticizes me unnecessarily, I will call her out on it." is much better.

Review these rules once every quarter and most importantly, assess honestly whether you met these rules, standards, and principles during that quarter. Write down how you did. Adjust these rules, standards, and principles accordingly for the next quarter. If you had a perfect score for any particular rule, then delete it and add a new one to work on.

Repeat this process every quarter for the next year.

Exercise 3: Brain Loops (daily)

On a daily basis, I want you to do this every morning when you are in the shower and then a few more times throughout the day.

You are going to come up with some "brain loops" which is basically a fancy way to say affirmations.

These brain loops are to be used to introduce and reinforce changes to your internal mental makeup based on the principles you learned in this book.

You are going to repeat these brain loops to yourself over and over (or if you have a place where you will not be heard, you can say these things out loud) for the time period.

When you are doing this, do it for 1-2 minutes and come up with a few brain loops that reflect the high value that you will now carry yourself with.

Here are some ideas that I use myself, but you are free to come up with your own:

"I love myself!"

"I value myself greatly!"

"The world is on my side!"

"No matter what, things will be okay."

Exercise 4: Grounding (as needed)

A high value man is a "grounded" man. What does that mean? We hear this term being thrown around from time to time, but I bet you may not know what it actually means.

Being "grounded" merely means in touch with your body, your balls, and being in the present.

Use this exercise whenever you feel like your mind is wandering, especially when it starts being anxious. This will help you with taking the sails out of any neurotic state of mind you're in, which as I wrote in an earlier chapter is one of the main characteristics of a low value man.

I did not invent grounding. There are many ways to ground yourself, but this is the way I do it. If this doesn't work for you, you can just search Google for grounding meditations or bioenergetics. This exercise is based on a type of "body-psychotherapy" that in simplistic terms uses your body to heal the mind.

By engaging in the grounding exercise with your body, you heal the mind, and that's why I personally find that this exercise helps me dispel any neurotic state of mind I may temporarily be in.

So, this is how I do grounding:

1. Stand with your feet on the ground.

2. Make sure your feet feel planted into the ground.

3. Now, imagine your mind going down your spine, sinking down all the way into your stomach.

4. Then, imagine roots coming from the base of your spine extending downward to your feet and then into the ground.

5. Imagine the roots pulling your entire body, including your brain and head, down into the earth.

6. Raise your arms over your head and keep them there.

7. Now, jump up and down just slightly, and while jumping up and down, let out a sound over and over. The sound I like to use is "Ha! Ha! Ha!" or "Huh! Huh! Huh!"

8. Do this for about a minute and you may stop and get on with your day.

Exercise 5: Visualization (as needed)

Whenever you encounter a situation where you need to be assertive, but you feel like you aren't in the right frame of mind, here is an exercise to help you immediately snap into a state of mind to be assertive.

1. Close your eyes and take a few deep breaths.

2. Imagine the problem you're facing in front of you and imagine yourself in the scene as if you're watching a movie of yourself.

3. Now, remove yourself from the scene and replace yourself with a small, helpless child.

4. Visualize the wrong that is occurring to the child in detail. In particular, visualize the perpetrator and the victim, especially the child victim's reactions. Feel the pain and hurt of the child.

5. Feel yourself getting angry, but in a grounded way. You are not going to lash out, but you will respond to this harm being perpetrated with conviction.

6. Then, take whatever action you were sheepish about taking.

(DOWNLOAD your SPECIAL BONUS *"The High Value Man Checklist"* at www.artofverbalwar.com/thvmbonus)

CHAPTER THREE
Conclusion

In writing **The High Value Man**, my challenge was to distill a lot of random thoughts I've had over the years (which were captured in scribbles in a notebook and in random places on my computer) along with many (yet again, random) things I've read from other authors during this same time into a workable whole, and then combining these things with my own real world experience.

At times, it felt like putting together an epic Lego creation but not knowing where all the pieces I needed were and not knowing exactly what the final creation looked like.

Unlike most of the books I've written to date, this one was the most challenging, but I suspect it will be the most rewarding.

As you continue your journey to transform yourself into a high value man, I suspect it will be a lot like my journey in writing this book.

It will undoubtedly be challenging.

It will require that you take a realistic and critical view of yourself.

And most of all, it will require that you gather hard work, patience, dedication, and commitment together. You have to have faith that your final "Lego creation" will come together. So, I urge you to remember this analogy and press through those times when you feel like you are making no progress.

At the beginning of this book, I talked about an inflection point and divergence in masculinity.

I hope this book has shown you another path for your masculinity other than the two paths that were previously known to you, and that you make a commitment to travel down this new path.

Making a commitment to transforming yourself is not for everyone. But, I hope you make this truly valuable commitment for yourself.

It is going through this challenging PROCESS that will make you who you will become.

The rewards to becoming a high value man are great and absolutely worthwhile.

I look forward to hearing about your transformation.

I'd love to hear about your journey down this path I have shown you, so if you would like to share your thoughts, successes, or even challenges with me, you may always reach me at info@artofverbalwar.com.

To winning,

Min Liu
San Francisco, CA

NEXT STEPS:

DOWNLOAD your **SPECIAL BONUS** *"The High Value Man Checklist"* (**$99 price, infinitely valuable**), containing a summary of ACTION STEPS to continue your transformation into a high value man at www.artofverbalwar.com/thvmbonus

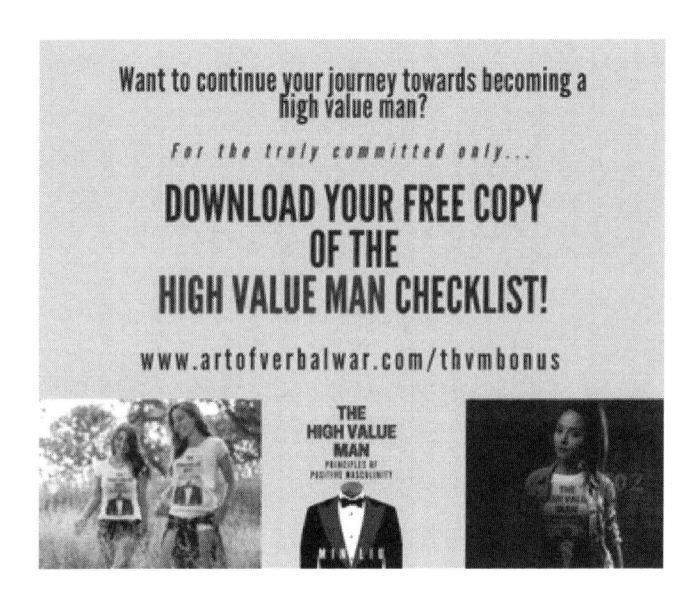

SUBSCRIBE to my YouTube channel, The Art of Verbal War, where people learn to EXCEL at verbal skills at www.youtube.com/artofverbalwar

READ MORE about verbal skills, power, persuasion, and influence at my blog at www.artofverbalwar.com/blog

SEND ME A MESSAGE at info@artofverbalwar.com

And, please REVIEW this book at Amazon.com so that readers like you may more easily discover The High Value Man!

About the Author

Min Liu is a corporate lawyer, Amazon #1 bestselling author, and the founder of The Art of Verbal War, where people learn to EXCEL in verbal skills.

Based in San Francisco, CA, Min's burning ambition is to teach like-minded people how to give their gifts and value to the world by helping them become EXCEPTIONAL in verbal skills, persuasion, influence and power.

In the words of his readers, he's the "older brother you've never had", and as a real-life big brother himself, his mission is to show you the ropes in all the things school never taught you.

He's especially aroused by basketball, meditation, reading books on psychology and inspirational people, people who are value givers, and most of all, constantly breaking out of his comfort zone and helping others break out of theirs. On the other hand, he despises value suckers, mediocre mindsets, and most of all, wearing sweaters.

Media, speaking, one-to-one coaching requests, or other inquiries can be sent to info@artofverbalwar.com.

Also by Min Liu

BOOKS

PEOPLE GAMES
The definitive guide to dealing with emotional manipulators (Kindle eBook)

THE KING'S MINDSET: TWENTY MINDSETS TO TRANSFORM ORDINARY MEN INTO KINGS
The ambitious man's "roadmap" to extraordinary success in life (Kindle eBook)

THE NEW ART OF BEING RIGHT: 38 WAYS TO WIN AN ARGUMENT IN TODAY'S WORLD
A reimagined version of Arthur Schopenhauer's "Art of Being Right", a playbook of strategies and tactics to help you win arguments and debates in today's complicated society
(Kindle eBook)

VOCAL SUPERSTAR: HOW TO DEVELOP A HIGH STATUS VOICE
Learn ten steps to develop a high status voice that will increase your influence and authority (Kindle eBook)

VERBAL SELF DEFENSE 101
An introduction to verbal self-defense (Kindle eBook)

To learn more about my books:
www.artofverbalwar.com/books

COURSES

VERBAL SELF DEFENSE FOR THE SOCIALLY INTELLIGENT
An online course about defending yourself from verbal bullying, attacks, and insults with wit and social intelligence

VERBAL DOMINATION
An online course about dominating and winning verbal confrontations

THE HARVEY SPECTER GUIDE
An online course about how to win big in life, inspired by the main character of the TV show "Suits"

THE HANK MOODY GUIDE TO WIT
An online course about how to ignite your wit and charm, inspired by the most charming man on TV, Hank Moody of "Californication"

MASTER OF METAPHOR
An online course about the verbal superpower of figurative language

To learn more about my courses:
www.artofverbalwar.com/courses

LAST CHANCE! DOWNLOAD your SPECIAL BONUS *"The High Value Man Checklist"* **at** www.artofverbalwar.com/thvmbonus

Printed in Great Britain
by Amazon